Cover illustration: Early Churchill Mk.IIs on exercises in 1942. Almost concealed by mud is the red/white/red Royal Armoured Corps recognition sign, whilst on the turret side is the sign for 'A' Squadron.

1. The first major use of the Churchill tank in action was with the First Army in Tunisia in January 1943, Churchill IIIs and IVs equipping the 25th Tank Brigade and later the 21st. In the hilly country of Tunisia the tank's lack of speed was no disadvantage and its heavy armour protection was a positive advantage, particularly when the formidable Tiger tank appeared. Here a Churchill III is seen in the classic infantry support role for which the vehicle was intended. The colour scheme is 'sand and spinach'.

TANKS ILLUSTRATED No 25

The Churchill Tank

CHRIS ELLIS

ARMS AND ARMOUR PRESS

Introduction

First published in Great Britain in 1987 by Arms and Armour Press Ltd., Link House, West Street, Poole, Dorset BH15 1LL.

Distributed in the USA by Sterling Publishing Co. Inc., 2 Park Avenue, New York, NY 10016.

Distributed in Australia by Capricorn Link (Australia) Pty. Ltd., P.O. Box 665, Lane Cove, New South Wales 2066.

© Arms and Armour Press Ltd., 1987.
All rights reserved. No part of this book may be reproduced or transmitted in any form or by any means, electronic or mechanical, including photocopying, recording or via an information storage and retrieval system, without permission in writing from the publisher.

British Library Cataloguing in Publication data
Ellis, Chris
The Churchill tank.—(Tanks illustrated; no. 25)
1. Churchill tank—History
I. Title II. Series
623.74'752   UG446.5

ISBN 0-85368-808-7

Edited and designed by Roger Chesneau; typeset by Typesetters (Birmingham) Ltd.; printed and bound in Italy in association with Keats European Production Ltd., London.

The Churchill tank was to the general public the best known British tank of the Second World War, an honour ensured by the fact that it was named after the British wartime Prime Minister, Winston Churchill, and was therefore easily remembered. It also shared its namesake's rugged, pugnacious character, and this was immediately obvious on sight. If ever a fighting vehicle were aptly named it was the Churchill tank. It enjoyed a similar longevity in the British Army also, from 1940 until 1965, which coincidentally was the year Winston Churchill died.

The genesis of the Churchill design was a requirement for a 'shelled area tank' called for in a War Office specification, the A20, drawn up on the outbreak of war in September 1939. The A20 design was in all essentials like an updated First World War tank – small wheels, all-round tracks, sponson-mounted armament, no track covers so that an 'unditching beam' could be attached, and an ability to cross trenches, surmount a 5ft obstacle running forward and a 4½ft obstacle in reverse. The Mechanization Board, responsible for actual design, met on 25 September 1939 and amended the concept to eliminate the side sponsons in favour of a turret and specified a 60mm minimum armour thickness to wtihstand the German 37mm anti-tank gun. Existing mechanical components were to be used and Harland and Wolff of Belfast were asked to do detailed design and production.

The A20 prototypes proved unsatisfactory in many ways, being too heavy and mechanically unsound, and able only to mount a 2pdr gun. By June 1940 the whole A20 project was cancelled, and Vauxhall Motors Ltd., of Luton, were asked to take over a revised design known as the A22. At this time Britain was desperately short of tanks (having fewer than 100 left after the Dunkirk evacuation), so top priority was given to the project at Winston Churchill's behest. The final design of the A22 was approved in late July 1940, the mock-up was ready in November, and the pilot model was on test by 12 December 1940. In view of Churchill's enthusiasm for the project, his was the name given to the new tank. There were, inevitably, teething troubles with the new design, and the first fourteen full production models were not delivered until June 1941. It was, however, Vauxhall's first tank design, in a purpose-built factory, and by the development standards of the time (and since) it was an astonishingly fast project, taking under a year from concept to production.

The Churchill tank was developed way beyond the most optimistic dreams of 1939 and 1940 and as both a fighting vehicle and a special purpose type it became a mainstay of the British Army through the Second World War and long beyond, as is shown (as chronologically as possible) in the pages of this book.

The author wishes to thank Peter Chamberlain, Hilary Doyle, Ron Rushton and Arthur North for assistance with photographs for this book, many of which came from their collections. Thanks are also due to Maire Collins, who typed the manuscript.

Chris Ellis

◀2
2. A top view of a Churchill II shows the rear mudguards (a retrospective modification), the exhaust pipe on the rear decking and the asymmetric shape of the early cast turret. Note also the tool stowage at the rear.

▲3

▲4 ▼5

3. The original design impression for the A20 'shelled area tank'. Note the overall track and the side-sponson machine-gun armament, like a First World War tank but with a Matilda tank turret and a 2pdr gun in the nose.
4. The A22 prototype built by Vauxhall Motors Ltd. on early running trials, still with a wooden simulated turret. The hull is shorter than that of the A20 and the side air intakes are bigger.
5. The Bedford Twin-Six engine which powered the Churchill. It was a simple but successful expedient produced by combining two 6-cylinder Bedford truck engines into a Flat-12 configuration. It developed 350hp.
6, 7. The first official photographs, by the Vauxhall company photographer, of an early-production Churchill, 9 May 1941. This original model had a 3in gun in the nose and fully exposed tracks. The turret gun was a 2pdr and the front horns were also fully exposed. Note also the masked 'black-out regulation' headlights.

▲8 ▼9

8. Prime Minister Winston Churchill took a great personal interest in the tank which bore his name. He was an early visitor to the factory in March 1941 and rode in T30971, the first production Mk.I. Here he leaves the front compartment, in white overalls and a black Tank Corps beret. This tank had a mild steel turret but the hull was fully armoured.

9. Prime Minister Churchill, watched by a Vauxhall engineer, talks by radio to Chief of the Imperial General Staff General Sir John Dill, who was riding in the second Churchill production tank. Note the case for signal flags on the turret side, a relic of prewar procedures, and the triangular plate warning that the turret of T30971 is mild steel, not armoured.

10. A good view of the 3in howitzer prominent in the nose of a Churchill Mk.I. This picture was taken in the planning stages, early in 1942, for the Dieppe raid, and the vehicle is checking clearances for leaving a landing craft. The side air intakes have been removed to reduce the vehicle's overall width.

11. A good close-up view of a Churchill Mk.I on delivery trials. Note the early type of armoured side intake, exposed track, and open front horns. The hawser stowage is well shown, as are the turret side box and the pistol port. On the front horns is the red/white/red recognition bar painted on most British tanks in the 1939–42 period, a revival of the First World War recognition marking.

10▲ 11▼

▲12
12. The Churchill II was similar to the Mk.I except that the 3in gun in the nose, which had a limited field of fire, was replaced by a 0.303 Besa machine gun. A production change on all Churchills was the addition of guards inside the front horns to stop dust and mud obscuring the view of the driver and front gunner. Triangular support plates in the corners held the dust guards in place, and are visible here. This Churchill II is leaving a 'Warflat' railway wagon and it can be seen that the side intakes have been removed to make it comply with loading gauge restrictions.

13. Vauxhall Motors acted as the 'parent' company for a group of firms which built Churchill tanks, among them Beyer-Peacock, Dennis, Harland & Wolff, Metro-Cammell and Leyland. Of these, the last two were the first to start production, in June and July 1941. This is a typical early-production Churchill Mk.II with the original style of side air intake. Because of the speed of the development programme early vehicles were sub-standard and unreliable, and the first thousand were 're-worked' in 1942–43 to bring them up to later standards.

▼13

14. The Churchill II CS (Close Support) was produced only in small numbers. It had a 2pdr gun in the nose and a 3in howitzer in the turret – a remnant of prewar policy where a few CS tanks were included in a regiment to provide smoke and HE fire.

15. From about March–April 1942 there was a change in the Churchill's appearance because of design improvements then put into effect. Full covers in three sections were fitted over the top run of the tracks, and there were new side intakes with top openings which made it possible to fit trunking to them for deep wading. Subsequently many of the old Churchill I and II vehicles were brought up to this standard, as typified by this Churchill Mk.I of the 12th Canadian Armoured Regiment in March 1942.

16. A summer 1942 exercise, with newly delivered Churchill IIs of an Army Tank Brigade causing a lot of interest among the accompanying infantry. Note the AA protection in the form of a Vickers 'K' machine gun on the PLM mount.

17. In April 1941, when the first German 50mm anti-tank guns were encountered in Libya, it was decided to fit the newly developed 6pdr gun in place of the 2pdr in Churchill tanks. This necessitated a larger turret, and a welded version was designed. The first Churchill III, as the 6pdr version was designated, was ready in February 1942. Essentially it had the Mk.II standard hull, with only a turret change, but note also in this photograph the extra fuel tanks at rear.

18. A good view of an early Churchill Mk.III in service, August 1942, showing the original style of air intake, spare track shoes, and the jettisonable spare fuel tanks at rear.

19. A cast turret was also produced for the 6pdr gun, and vehicles so fitted were known as Churchill IVs, but, owing to limited manufacturing capacity, these variants were less numerous. The cast turret was nominally better protected than the welded Mk.III turret.

▲16 ▼17

20. The Churchill Mk.III was first used in action in the Dieppe landing of 19 August 1942, with the 14th (Canadian) Calgary Tank Regiment. Some Mk.Is were also used, and of 30 vehicles sent on the raid, 28 actually landed from LCTs. Three vehicles were fitted as flame-throwers, and leading vehicles carried primitive bolts of chestnut paling to lay a carpet on the shingle. The use of tanks was abortive, however, and all were knocked out, most on the beach. Here German soldiers examine a knocked-out vehicle on the Dieppe seafront.

21. The Germans captured enough material at Dieppe to make up an operative Churchill III for evaluation, and here it is in German service. The problems of landing AFVs over an open beach led to the development of special-purpose tanks, many based on the Churchill, which were used in the Normandy landings of June 1944. Note on this vehicle the early form of wading gear and an indication of the armour thicknesses.

▲20 ▼21

22. Two Churchill IIs were shipped to Egypt for tropical testing in December 1941, and three Mk.IIIs were sent out in the autumn of 1942. They were present at the Battle of Alamein in October 1942 but were used only in a minor role as Brigade HQ vehicles. One was knocked out and two broke down, but they were publicized for propaganda purposes at the time with the inference that they played a more important part.

23. The early Churchill IIIs were still in service long after production standards were improved and hull changes introduced. An early Churchill III with a Mk.II-standard hull is seen here taking part in Exercise 'Spartan' in March 1943. It is from the 12th Canadian Army Tank Regiment (Three Rivers Regiment) and displays the red/white/red recognition bars on nose and side, the 'C' squadron symbol on the turret side, and Canadian formation signs on the front.

▲24

24. A Churchill II named 'Artichoke', of the 14th Canadian Army Tank Regiment (Calgary Regiment), literally brings the house down during a final training session in England immediately before the Dieppe raid of 19 August 1942. Note the early-type side intakes and the mesh stowage bin on the turret rear, the latter not a common fitting.

25. The Churchill V was a new close-support tank produced in 1943, intended to replace the remaining Mk.Is which had been retained for close-support purposes by virtue of their 3in howitzer armament. The Churchill V had a 95mm tank howitzer Mk.I with coaxial 7.92mm machine gun. It was built in small numbers and was essentially a Mk.IV with a cast turret but with altered armament.

▼25

26. Cleaning the main armament of a Churchill IV in Italy. This is the 6pdr Mk. 5 gun, distinguished by the prominent counterweight on the muzzle.

27. The other type of 6pdr gun fitted to the Churchill III and IV was the Mk.3, which had a less tapered barrel and lacked the counterweight. This view shows a Churchill III with 6pdr Mk.3 gun being loaded on to a 'Warflat' with the side intakes removed to prevent overhang; they can be seen placed on the engine decking. Width restrictions to suit the railway loading gauge were a limitation on up-gunning the Churchill to take a weapon larger than 75mm.

26▲ 27▼

▲28 ▼29

28, 29. The British 2pdr and 6pdr tank guns fired only armour-piercing rounds. The American Grant and Sherman tanks entering British service in 1942 had 75mm guns, able to fire both armour-piercing and high-explosive rounds, thus enabling them to engage both 'soft' and 'hard' targets and eliminating the need for separate close-support tanks; in addition, German tanks now had 75mm guns and there was obviously an urgent need for 75mm guns fitted to British tanks. The first fitting was a field modification in Tunisia by the 21st Tank Brigade, whose workshops took American M3 75mm guns and mantlets from wrecked Sherman tanks and mounted them in 120 Churchill IVs. This type was called Churchill IV (NA 75). With a capacity for 75 rounds, it was a successful conversion, and the type was used in Italy until the end of the war. Here can be seen a squadron of Churchill IV (NA 75)s providing support fire, and details of the mantlet.

30. Vickers developed a British version of the 75mm gun, essentially by boring out a 6pdr. The design work started in December 1942, but various difficulties prevented production guns from being ready until February 1944. The British gun could fire US ammunition. Some Mk.IIIs were fitted with 75mm guns, but most were fitted to Mk.IVs and this type, with cast turret, was designated Churchill VI. Capacity was 84 rounds. Note the 'wading height' marked on this early Mk.IV.

31. Though a heavily armoured vehicle by the standards of other British tanks, the Churchill was no match for the German 88mm gun. Here a Churchill IV has been penetrated and knocked out by accurate 88mm fire. This vehicle, photographed in November 1943, is one of a number of Churchill IIs, IIIs, and IVs supplied to the Soviet Union under Lend-Lease in 1942–43. Note the red star marking and worn call-sign (208) apparently painted in whitewash.

32. A scene at the Vauxhall motor works, Luton, 1943, where new Mk.IIIs are undergoing final inspection and checks before delivery. They have just completed immersion tests – half an hour in water with the hull submerged, during which time no water must have penetrated. Note the Army personnel testing the communications.

▲33 ▼34

33. Engine access on the Churchill was straightforward: the covers lifted for access and were stayed in an upright position. Note also in this photograph the exhaust manifolds and the tools stowed behind them.

34. While the 75mm gun was being developed it was decided to improve the basic Churchill design to incorporate all the lessons learnt in war. The outcome was the A22F, designated Churchill VII. It looked externally similar to previous marks, but the hull was now made in a single thickness of armour plate – up to 6in – in place of the layers of appliqué armour on earlier marks. Circular side escape hatches replaced the square ones (reducing armour weakness at this point) and the driver's vision port was also made circular. The turret was completely redesigned, with cast sides, a welded roof, and a vision cupola for the commander.

35. The Churchill VII was a great improvement, but the redesign increased the weight of the tank to 40 tons and reduced its top speed to 13mph. In service the Mk.VII was soon encumbered with gear and, like the earlier marks, had parts of the track covers removed. The Churchill VII was available just in time to participate in the Normandy campaign. This Mk.VII is from the 6th Guards Tank Brigade, 3rd Scots Guards, Flank Company, and it is crossing a crater by means of a Churchill tank bridge. Note the spare track shoes, giving extra protection, and the missing track guard centre-section. The 75mm gun and circular escape doors identify the vehicle as a Mk.VII.

▲36

36. Churchill VIIs were thinly spread at first. In this 3rd Scots Guards troop, the nearest tank is a Mk.VII, and the next two are Mk.VIs. Infantry of the 15th Scottish Division are here waiting to move forward with tank support in Holland, late 1944.

37. The Churchill Mk. VIII was the close-support variant of the Mk.VII. It was identical except for a 95mm howitzer main armament which replaced the 75mm gun, and it was easily

distinguished by the short barrel with counterweight. The Mk.VIII was only produced in small numbers because the 75mm gun of the Mk.VII could fire both HE (high-explosive) and AP (armour-piercing) rounds, so rendering obsolete the requirement for a separate tank to fire HE. This view shows well the extra thickness of the side armour on the Mk.VII/VIII hull and a slightly different plate shape.

▼37

38. A rear view of the Churchill VIII shows features of the Mk.VII/VIII, including the rear towing eye, racks for fuel cans on the rear mudguards, the telephone box for accompanying infantry (on rear plate), heavier exhaust pipes, spare shoes stowed on the side, the new heavy cast turret, and the commander's vision cupola.

39. In 1944–45 the Churchill Mk.VI, with 75mm gun, was in service with the Mk.VII, some units mixing the marks because the Mk.VII was initially in short supply. This Mk.VI shows clearly how the side armour was made up of layers to give adequate thickness; in the Mk.VII the sides were one-piece plates. Prior to the introduction of the Churchill Mk.VII, the side escape doors were square.

▲40 ▼41

40. An early version of Churchill flame-thrower tank was the Oke, which consisted of a standard Ronson flame device (normally man- or carrier-transported) with the fuel tank fitted on the hull rear and the flame projector pipe carried in the side sponson to fire forward from the front horn. This had insufficient capacity so the Petroleum Warfare Department, the responsible body, developed a gas-operated flame projector in which the fuel for flaming was carried in an armoured, wheeled, 400-gallon trailer. This had a quick-release, pivoted coupling to the vehicle and the trailer could be jettisoned when empty, leaving the Churchill to fight as a normal gun tank. Called the Crocodile, it was attached to specially modified Churchill VIIs.

41. The flame projector pipe was carried from the trailer connection under the belly of the tank and exhausted through the hull machine-gun port. The flame projector (and the co-driver/flame operator) can be clearly seen in this view of Churchill Crocodiles moving in to attack Brest. Apart from the flame projector and trailer coupling there were no other changes to the tank.

42. The experience of the abortive Dieppe raid in August 1942 showed that specialized armoured vehicles would be invaluable for landings over defended beaches, to assist in overcoming obstacles, climbing sea walls, placing demolition charges, etc. This led to the formation of the 79th Armoured Division, charged with developing suitable vehicles. One of the most important types produced was the Armoured Vehicle Royal Engineers (AVRE), and the Churchill tank proved to be the best type for the job. The short Petard mortar instead of a gun was its most distinctive feature, but it also had bolt positions on the side to which could be anchored a dozer blade and other attachments. The prominent bolts are visible here.

42 ▼

▲43 ▼44

43. This view of a fully equipped Churchill AVRE shows the guide rails and pivot plate for a dozer blade attached to the side bolts, spare track shoes and 'spuds' carried on the hull and turret and, in this case, two spare bogie assemblies on the rear track covers.

44. A rear view of the same Churchill AVRE conversion shows the tow hook, infantry telephone box, racks for water cans, and two spare bogie assemblies on the rear track covers. The AVRE went into production early in 1944, converted from the Churchill III or IV (mostly the latter). Alterations included new internal stowage for demolition stores and the Petard ammunition, and a sliding hatch in place of the co-driver's flaps, to facilitate vertical loading of the Petard mortar from inside the forward compartment.

45. A close-up view of the Petard mortar on an AVRE. It was attached to the original 6pdr mantlet of the Churchill IV and the breech opened by tilting the barrel vertically, loading being effected from the sliding front hatch in the hull. The co-driver acted as loader.

46. The 'Flying Dustbin' projectile that was fired by the 29cm Petard mortar. It weighed 40lb, could be fired at the rate of 2–3 rounds per minute, and had an effective range of 80yds.

45▲ 46▼

29

▲47

47. An AVRE running at full speed during the attack on Colombelle, 19 July 1944. On this vehicle all the track covers have been removed (rare at this late stage of the war), and an old ammunition box has been attached to the hull side as an extra stowage aid.

48. The AVRE could also carry a fascine (bundle of wood) on its nose to fill craters and anti-tank ditches. A wooden cradle supported the fascine (and can be seen here), hawsers held the fascine in place until release was required, and a slip line allowed release to be carried out remotely from under cover in the turret. As the commander's view was obscured by the fascine, he often conned the vehicle from atop the fascine as shown here. Note also the flaps over the side intake and the extra track shoes.

▼48

49. All AVREs could, if required, carry a Standard Box Girder (SBG) bridge, as shown here. It was supported by sheer legs and was topped up on the nose by a winch in a carrier frame which was attached on the AVRE's rear deck. The SBG could be slipped by a rope from the turret. This AVRE with SBG is at Normandy, with a standard AVRE at right.

50. One of the most widely used attachments for the AVRE was the carpet-laying device which placed a canvas trackway over soft sand or pebbles to help AFVs leaving landing craft. The first type was the TLC carpet layer, a small roll on dozer-blade arms. There was a twin-bobbin version, but the Bobbin Mk.I was the most effective. Note how the arms are attached to the standard AVRE hull fittings.

▲51 ▼52

51. There were numerous mine ploughs attached to AVREs, some used operationally in small numbers and others being no more than experimental. This is the Farmer Deck Plough, intended for mine clearing, again carried on arms which fitted the AVRE side stations.

52. The CIRD (Canadian Indestructible Roller Device) was a Canadian Army idea. Several models were developed, with 16in, 18in and 21in rollers. This is the first model, CIRD 16in.

53. The CIRD 18in saw limited production. The size indicates the diameter of the rollers which acted as the mine exploders.

54. The Farmer Track plough was designed to uncover the mines and it also had rollers which could explode mines that were not pushed aside. It was developed in late 1943. The vehicle is an AVRE with a standard turret and 6pdr gun.

55. The standard wading kit for all Churchills with the later side intake. Trunking is fitted to the top of the intakes and for deep wading waterproof sealing was put round the turret base and over other openings. This vehicle has a damaged rear mudguard.

53▲

54▲   55▼

56. An armoured recovery vehicle conversion of the Churchill was developed in early 1942. It was essentially a Churchill I or II with the turret and armament removed. It had a rear drawbar for towing, and a jib which was normally carried stowed on the vehicle; timber baulks, stakes and haswers were stowed on the hull. Here a Churchill ARV Mk.I of the 6th Guards Tank Brigade is being loaded on to a Diamond T tank transporter during a demonstration for HM King George VI in 1943.
57. A Churchill ARV Mk.I in service, removing a sprocket wheel from a damaged Churchill IV during exercises in Britain in 1943. The jib and chain block at the head had a lifting capacity of 3 tons.
58. The Churchill ARV Mk.II was more lavishly equipped. Converted from the Churchill IV, it had a fixed turret-like structure with a dummy gun, a fixed rear jib of 15 tons capacity, a winch, a dismountable forward or rear jib, and a spade-like earth anchor. This example is crossing the Rhine by pontoon, 24 March 1945.

▲59
59. The Churchill BARV (Beach Armoured Recovery Vehicle), of which only a very few were built. It had trunking for wading and a superstructure similar to the ARV Mk.II but without the jib or dummy gun. The Sherman proved to be better for beach recovery work and production was switched in favour of the Sherman BARV. This vehicle is towing Universal Carriers ashore during a pre-'Overlord' exercise in 1944.
60. One of the major problems when landing on an open beach was getting over the sea wall – this was a stumbling point at Dieppe. The Churchill Ark Mk.I was the answer to the problem. It was a turretless Churchill III or IV with trackways of metal and timber construction above the tracks and with short hinged ramps at each end. A ratchet attachment under the front ramps allowed the tracks to bounce the ramps against the sea wall. Fifty Ark Is were produced in February 1944 for use in the Normandy landings. Here the sea wall incline is rehearsed.
61. Following tanks could pass over the Ark and thus on to the sea wall. The Ark was also used to span rivers, or anti-tank ditches, and was considered expendable if necessary. The Ark Mk.II was similar to the Mk.Is (which were later converted to Mk.IIs), but had both a 2ft and a 4ft trackway, thus allowing smaller vehicles to pass.
62. The Ark Mk.II (Italian Pattern) was developed by British forces in Italy and used its own tracks as a trackway. The ramps were American Treadway tracks and came in two lengths, with king-posts to hold them up. Note the extra width on the left-hand side ramps to help smaller vehicles. Ark was an 'acronym' for 'armoured ramp carrier'.
63. (Next spread) The Churchill Bridgelayer was a major type which was developed in 1944 in time for the Normandy invasion. Converted from a gun tank, it had a hydraulic drive carried in the front turret space which operated a pivot arm to launch a 30ft tank bridge No. 2, the latter borne on supports above the hull. This overhead view shows the hydraulic launch arm, with rollers for positioning.

▼60

61 ▲ 62 ▼

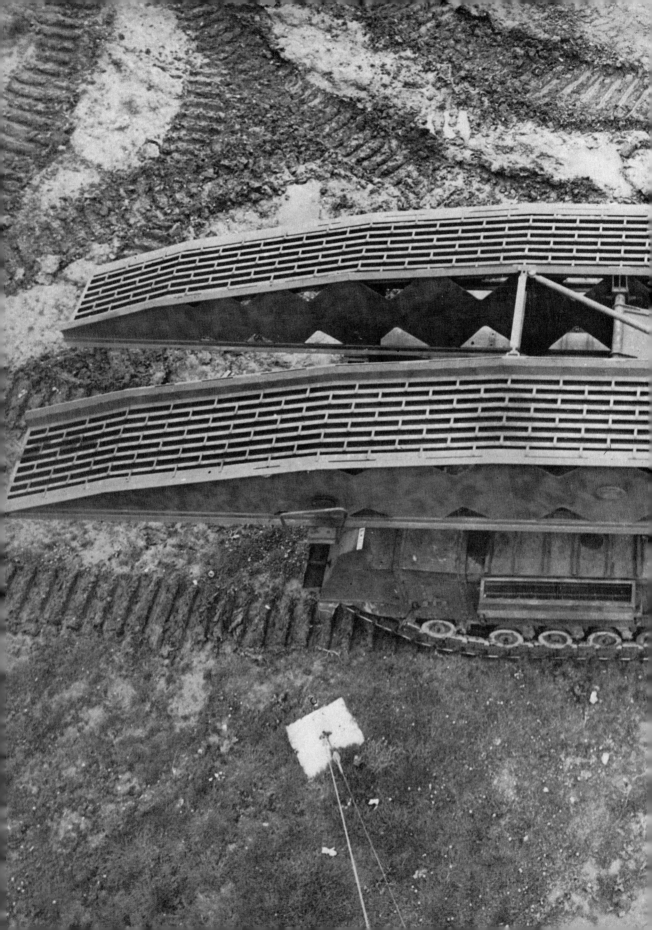

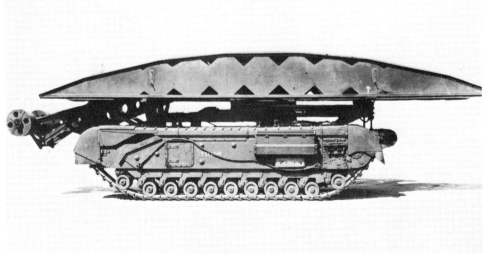

64. The Bridgelayer conversion was based on a Churchill III or IV. The bridge could carry a 60ft tank or Class 40 wheeled vehicles, and the driver operated the laying mechanism; the only other crew member was the commander.

65. A side view of the Churchill Bridgelayer shows well the size of the bridge – which was large for its day. The Bridgelayer remained in service until well into the 1960s.

66. The Churchill AVRE was used on occasion to propel a number of mobile bridges for spanning rivers and other obstacles. Mostly these were Bailey bridges, specially built for the task in hand. The Mobile Bailey was the most frequently used, notably to cross the Senio River in Italy in April 1945. It would span a 70–80ft gap, was assembled off site on tracked dollies, and was pushed into position by the AVRE. The early version had a turretless Churchill as the carrier vehicle instead of the tracked dollies.

67. A new Churchill Mk.III, built by Dennis Brothers, undergoes running trials at the company's works, with a wood canopy substituting for the turret which will be fitted later. Note the extra fuel tank on the rear.

▲68
68. A Churchill VI line-up at the Dennis works. These are rebuilt Mk.IVs; the 75mm guns still await their counterweights, and the side intakes are not yet fitted. Note how the front dust guard can be hinged upwards to clear mud.

69. Tanks are carried by transporters where possible, to avoid costly wear and tear. Here a Churchill VI is unloaded from a trailer at Arezzo, Italy, in 1944, where a complete Churchill regiment has been brought up to support an infantry advance.

▼69

70. The Carrier Churchill, 3-inch Gun, Mk.I was an attempt to put a high-velocity gun on a tank chassis to give fire power to match the German 88mm gun. No British tank was big enough to take a gun heavier than the 6pdr, and it was therefore decided to fit a gun in a limited traverse mount in the hull of a Churchill tank. Surplus 3in (12½pdr) AA guns were adapted for the purpose, and the project for 50 vehicles was initiated in September 1941, the pilot model being ready in February 1942. Trials were successful, but these vehicles never saw combat: by 1943 the Challenger 17pdr gun tank was in development and was thought to be a better alternative.

71. The use of the Churchill 3in Gun Carrier was delayed because of indecision about its mode of employment and whether it was a tank or a self-propelled gun. Some were converted in 1943–44 to carry Snake mine clearing equipment, but these vehicles (which had the gun removed) were used only for trials.

▲72
72. By 1945 a rework programme was started to bring the Churchill Mk.III, IV, V and VI as close as possible to Mk.VII standard. Modifications included a reinforced roof and appliqué armour on the sides, a vision cupola and a vane sight in the turret, and a 75mm gun, and some vehicles were fitted with the Mk.VII turret. The reworked vehicles were designated Mk.IX (III or IV with Mk.VII turret), Mk.IX LT (light turret, the original type retained), Mk.X (Mk.IV with Mk.VII turret), Mk.X LT (Mk.VI with original turret), or Mk.XI (Mk.V brought to Mk.VIII standard). This is a Mk.X LT, photographed in 1946.
▼73

73. In the immediate postwar years there were plenty of spare Mk.VII Churchills available, so a new batch of Churchill Bridgelayers, based on the Mk.VII chassis, was authorized. Note here the tail lights, the rear stowage box and the covers over the side intakes.
74. A good detail view of the cradle and launching ramp of a postwar Churchill VII Bridgelayer, looking forward from the rear. Note the small cupola for the commander alongside the main hydraulic ramp.

74▶

75. The Irish Army were the last users of Churchill tanks in any quantity. This is a Mk.X LT in 1969, showing the turret vision cupola and much extra appliqué armour on the sides. Visible too, is the stronger suspension, based on the Mk.VII type.

▲76 ▼77

76, 77. The Churchill APC (FV 3904) was an early postwar adaptation of the Churchill VII to produce an armoured personnel carrier. The turret and associated gear were removed, bench seats for an infantry section were provided, and appropriate radio equipment was installed. The hull machine gun was retained and steps and treadways for infantry egress over the side of the vehicle were provided. Very few of the vehicles were produced, and they were mainly used by the School of Infantry.

78, 79. A wholly postwar (1950s) development was the Churchill Flail (FV 3902), also known as the Toad. It was a replacement for the Sherman Crab in the mine-clearing role. Chains on a rotor drum beat the ground ahead of the tank, the rotor being driven by an engine in the fighting compartment. These views show the rotor arm in the lowered and raised positions, but for travelling the arm was pivoted back over the rear. The assistance of two other tanks with hawsers was necessary to move the flail rotor from and to this position. The new superstructure was heavily armoured in front, to protect the vehicle from exploding mines. The Toad weighed 56 tons and had a crew of two (driver and commander), who could operate remotely the Whyman lane marker, the prominent device on the back which fired markers into the ground to indicate swept paths. Toads were withdrawn in 1965.

▲80 ▼81

80, 81. The Churchill Mk.VII, plus some Crocodiles and Churchill Bridgelayers, saw action again in 1950–51 in the Korean War. The 7th Royal Tank Regiment was one of the British units sent to join the United Nations forces aiding South Korea, and the tanks are shown here guarding the banks of the Han River near Seoul, the South Korean capital. An extra stowage bin is a local addition to the turret side of the tank on the right.

82, 83. The most important postwar development was the Churchill VII AVRE (FV 3903), which had a 165mm (6.5in) low-velocity gun in place of the old Petard mortar of the original wartime AVRE. A close view of the gun is shown in photograph 82, the complete vehicle being depicted overleaf in photo 83. There were also twelve smoke dischargers on the turret. The vehicle could be fitted with a derrick, a dozer-blade, a fascine cradle or a 30ft Tank Bridge No.3 carried on a front frame like the wartime AVRE SBG. The Churchill AVRE VII entered service in 1954 and was withdrawn from use in 1965 when it was replaced by the Centurion AVRE.

82▼

▲84 ▼85

84, 85. The Churchill Twin Ark (FV 3901) was a postwar version of the Ark designed to carry the super-heavy Conqueror tank across gaps. The width and weight of the Conqueror necessitated two Arks side by side, each providing one trackway for the crossing. The ramps differed from the wartime Ark in that they were of the folding type which were opened out by a system of hawsers and pulleys. As they approached the crossing site the two vehicles coupled themselves together and were interchangeable from side to side. The system was also known as the 'linked dog'. A single vehicle could be used to provide an Ark for soft skins or light tracked vehicles. The Twin Ark saw service from 1954 to 1965.

86, 87. Another postwar development for the AVRE was the Moyers Fascine Launcher. This equipment enabled an AVRE to carry and launch its fascine from above the turret and so overcome the problem of the poor visibility for the commander and driver experienced when the fascine was carried in the usual position on the nose; it also allowed the armament to be used, which was hitherto not possible, and a much larger bundle could be transported, as typified by this fabricated dumb-bell fascine. The carrying position is shown in photograph 86, whilst photograph 87 shows the turret traversed for launching, with the commander lowering the ramp and releasing the fascine. Despite its clever design this system was never adopted.

86 ▲   87 ▼

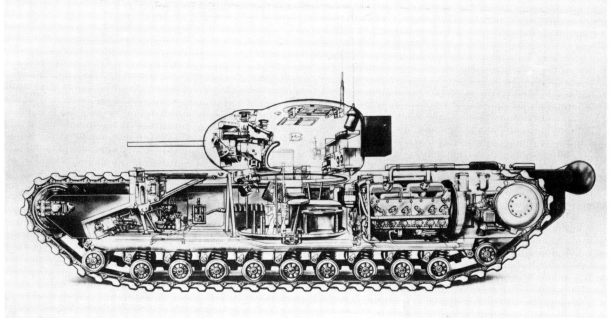

CHURCHILL,
CUT AWAY VIEW

▲88 ▼89

88. A cross-section through a Churchill IV shows well the precise layout of the vehicle. From left to right may be seen the front horns with idler wheel and track tensioning screw, and the driver's and co-driver's positions in the front compartment with their escape door adjacent in the sponson side. Ready-use ammunition is seen in the turret cage (it was also carried in bins each side of the turret), whilst in the turret itself can be seen the hand traverse handle and the standard No.19 wireless set. Just above the breech of the 6pdr gun is the roof-mounted bomb-thrower. The Bedford Twin-Six motors drive the rear sprockets via a gearbox. On the back of this example can be seen the infrequently carried extra fuel tank.

89. All this equipment and ordnance was fitted in a Churchill III or IV to make it combat-worthy. Note the small arms for the crew (Thompson machine guns in this case), the Bren gun, and the Besa machine guns for nose and turret.

90. Ammunitioning a Churchill VII with 75mm rounds, France, summer 1944. The tank commander is passing shells to the turret gunner for stowage in the 'ready use' racks in the turret cage. Note the Rexine-covered sponge rubber padding on the inside of the gunner's hatch flaps.

▲91 ▼92
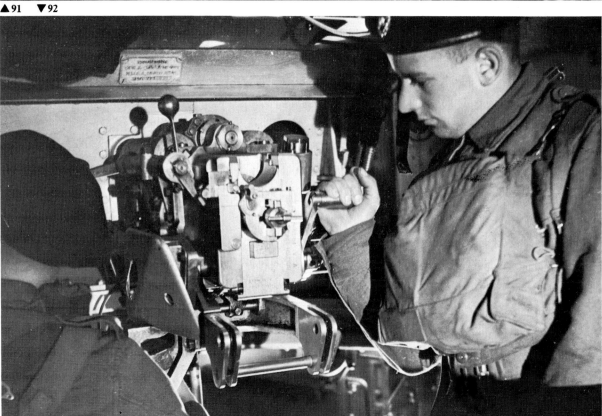

91. The turret gunner manning the 2pdr gun of a Churchill II (right), with the wireless operator (left) working the breech. The wireless operator was also the gun loader when in action.
92. Another view inside the Churchill II turret, with the gunner at his sights (left) and the wireless operator working the breech (right) in his capacity as loader.
93. The driver of a Churchill II at his controls on the right side of the front compartment.
94. The gunner with his sighting telescope set to the left of the gun.

93▲ 94▼

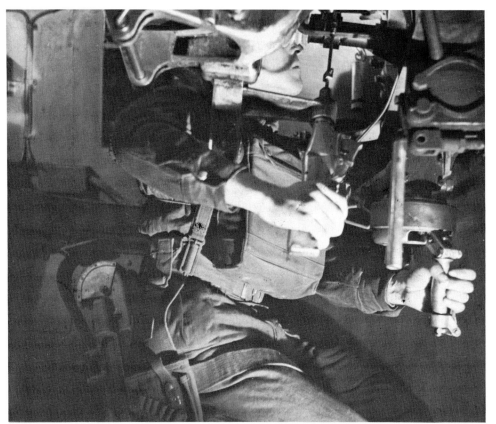

95. The gunner in a Churchill II demonstrates his action position. He is perched on his seat to the left of the breech, his left hand is on the traversing control handle (for training the turret and gun) and his right hand is on the trigger handle.

96. A front view showing the driver's front vision flap and vision block, the front 7.92mm Besa machine gun with its coaxial sight, and the periscopes for driver and hull gunner. In the turret mantlet can be seen (from left to right) the 6pdr gun, the coaxial 7.92 Besa machine gun, and the gun sight aperture. The loader's periscope is to the left in the turret roof.

97. The commander (right) and wireless operator in the turret of a Churchill II. In the internal mantlet can be seen (left to right) the coaxial 7.92mm machine gun, the 2pdr gun and the aperture for the gun sight. The vanes above the sight on the turret roof are to give the commander the line of the turret when observing from the periscope in front of the commander's hatch.

▲95 ▼96

97▶

▲98 ▼99

98. A close-up view of a Churchill IV turret, showing the vane sight for the commander, the commander's revolving cupola behind it, the periscope for the loader, the 6pdr gun and the coaxial Besa 7.92mm machine gun. Note, on the side, the canvas case for the commander's signal flags.

99. Maintenance work on Churchill VIIs of the 6th Guards Tank Brigade, Germany, March 1945; the ease of access to the engine compartment is apparent. The nearest tank has its turret sides festooned with track shoes, but note that these are from Sherman tanks, not Churchills. Old ammunition boxes are fixed to the turret stowage box, and the crew's helmets are attached to the hatch flaps.

100. Another example of the extensive use of track shoes for extra protection. On this Churchill IV old Sherman tracks are positioned on the turret front, but Churchill track shoes may be seen on the nose sides and also to the rear of the turret.

101. A close-up view of the Petard mortar on an AVRE IV. Note the protective cover over the coaxial machine guns and the rather crudely cut away turret front.

100 ▲ 101 ▼

▲102
102. A Churchill IV (NA 75) in Italy, early March 1945. The external mantlet is a complete lift from a wrecked Sherman tank to produce a Churchill with a 75mm gun some time before an 'official' model was developed.

103. Back to peacetime smartness: Churchill VIIs, freshly painted and all properly stowed, take part in HM The King's Birthday Parade on 2 June 1945 at Kiel.

▼103